# BRITISH PELAGIC TUNICATES

A NEW SERIES
Synopses of the British Fauna
No. 20
Edited by Doris M. Kermack and R.S.K. Barnes

# BRITISH PELAGIC TUNICATES

Keys and notes for the identification of the species

## J. H. FRASER

*The Marine Laboratory, Aberdeen*

1981
*Published for*
The Linnean Society of London
*and*
The Estuarine and Brackish-Water Sciences Association
*by*
Cambridge University Press
*Cambridge*
London  New York  New Rochelle
Melbourne  Sydney

Published by the Press Syndicate of the University of Cambridge
The Pitt Building, Trumpington Street, Cambridge CB2 1RP
32 East 57th Street, New York, NY 10022, USA
296 Beaconsfield Parade, Middle Park, Melbourne 3206, Australia

First published 1982

Printed in Malta by Interprint Limited


*British Library Cataloguing in Publication Data*
Fraser, J. H.
British pelagic tunicates. – (Synopses of the British fauna; 20)
1. Ascidiacea – Great Britain
I. Title        II. Linnean Society of London
III. Estuarine and Brackish-water Sciences Association        IV. Series
596.2   QL613   80-42174

ISBN 0 521 28367 1

# A Synopsis of
# British Pelagic Tunicates

## J. H. FRASER

*The Marine Laboratory, Aberdeen*

## Contents

|  | *page* |
|---|---|
| *Foreword* | vii |
| **Introduction** | 1 |
| *Practical methods and preservation* | 1 |
| Key to classes of British pelagic tunicates | 2 |
| Classification | 3 |
| **Class Appendicularia** | 4 |
| General biology | 4 |
| General structure and feeding | 4 |
| Life history | 6 |
| Locomotion | 6 |
| Ecology | 6 |
| Distribution | 7 |
| Systematic section | 8 |
| Key to British appendicularians | 8 |
| **Class Thaliacea** | 15 |
| *Order Doliolida* | 15 |
| General biology | 15 |
| General structure and feeding | 15 |
| Life history | 16 |
| Ecology | 18 |
| Distribution | 18 |
| Identification | 19 |
| Systematic section | 21 |
| Key to British doliolids | 21 |
| *Order Salpida* | 24 |
| General Biology | 24 |
| General structure and feeding | 24 |
| Life history | 26 |
| Ecology | 26 |
| Systematic section | 28 |
| Key to British salps | 29 |
| Key to aggregate forms | 29 |

Key to solitary forms                    30
*Order Pyrosomida*                       45
General biology                          45
  General structure           45
  Life history                46
  Ecology                      46
Systematic section                       47
Key to British pyrosomes                 47
*Acknowledgements*                       50
*Glossary*                               51
*References*                             53
*Species index*                          56

# Foreword

*Synopsis* No. 1 was devoted to those tunicates in which the adults are sessile (the class Ascidiacea); this one treats the two remaining tunicate classes (the Appendicularia and Thaliacea) which comprise entirely pelagic forms. All tunicates, as their name suggests, are invested by a test or tunic – often largely formed by a type of cellulose – and in many ascidians this test is tough and cartilaginous. In many of the pelagic species, however, it is gelatinous and translucent; in the extreme case of the appendicularians, cellulose is absent from the test and this entirely gelatinous structure, which may be replaced several times in one day, forms the 'house'. Nevertheless, as in other tunicates, the test is intimately associated with the feeding system.

The pelagic species may be extremely abundant. The author of this *Synopsis* quotes a case in which their numbers were sufficient to clog the cooling-water intake of a ship and hence stop it. Their numbers can also build up very rapidly as their phytoplankton food blooms: appendicularians may have an interval of only three days between generations; and some salps can double their numbers in a single day. Appendicularians, in turn, are of particular importance as the food of the young stages of some commercially important fish, such as the plaice.

The editors are indebted to Dr Fraser for producing this concise account of the pelagic tunicates to be found in British waters, and believe that the *Synopsis* will not only serve a strictly utilitarian purpose in respect of their identification but introduce these relatives of the vertebrates to a much wider public.

R. S. K. Barnes
Estuarine and Brackish-water
Sciences Association

Doris M. Kermack
Linnean Society

# Introduction

The pelagic tunicates of British and other waters are divided amongst two classes, the Appendicularia and the Thaliacea. Only the larvae of the third tunicate class, the Ascidiacea, are pelagic (Millar, 1970 – *Synopsis* No. 1).

A number of different names are used for the class Appendicularia: Larvacea, Copelata and Perennichordata. Whilst following Büchmann (1969) and Büchmann and Kapp (1975), the term Appendicularia is used in this *Synopsis*, the terms Larvacea and Copelata also have their advocates, although Perennichordata is now rarely used. Copelata is frequently used as the name of the single order contained within the class.

The Thaliacea are further divided into three orders, Doliolida, Salpida and Pyrosomida. As the general structure, life history and ecology of all four orders of pelagic tunicates are fundamentally different, they will be dealt with separately below. A very full bibliography of works on pelagic tunicates is given to 1973 by van Waard-Pouw and van Soest (1973).

## *Practical methods and preservation*

As all the British species are marine and planktonic, the most widely used method of capture is the plankton net in its various forms (Fraser, 1968). They are adequately preserved for normal taxonomic purposes in formalin – 4 % formalin or 2 % formaldehyde, preferably in sea water. As there are no calcareous parts, formalin will suffice without special buffering. If the plankton collection contains many salps, however, the high proportion of jelly must be considered as sea water and allowed for when making up the preservative. Labelling should be on good quality paper, written with a soft pencil or Indian ink, and placed inside the collection jar. A duplicate label can be affixed for convenience outside the jar. The information on the label should include collection serial number, date, place of capture, depth, and type of net used. Collections with much jelly, houses of *Oikopleura* (see p. 4), chains of diatoms, etc. will retain a much greater ratio of small forms in a relatively coarse mesh, and the small size and delicate structure of Appendicularians in particular makes them easy to miss in sorting the collections (Büchmann, 1973).

# Key to classes of British pelagic tunicates

1. Individuals with a pronounced tail* ...................... APPENDICULARIA (p. 4)

   Individuals with no pronounced tail ............................(THALIACEA p. 15) **2**

2. Tubular colonies comprising many zooids ............... PYROSOMIDA (p.45)

   Individuals singly or in chains............................................................................ **3**

3. Individuals with muscle bands in complete loops encircling the body
                                                          DOLIOLIDA (p. 15)

   Individuals with muscle bands not in complete loops, often branching and anastomosing............................................................................SALPIDA (p. 24)

---

* Take care not to confuse appendicularians with the 'tadpole' larval stage of ascidians.

# British pelagic tunicates

# Classification

Class Appendicularia
  Order Copelata
    Family Oikopleuridae
      *Oikopleura dioica* Fol, 1872
      *O. labradoriensis* Lohmann, 1896
      *O. fusiformis* Fol, 1872
    Family Fritillaridae
      *Fritillaria borealis* Lohmann, 1896
Class Thaliacea
  Order Doliolida
    Family Doliolidae
      *Doliolum* (*Dolioletta*) *gegenbauri* Uljanin, 1884
      *D.* (*Doliolum*) *nationalis* Borgert, 1894
      *D.* (*Doliolina*) *mülleri* Krohn, 1852
  Order Salpida
    Family Salpidae
      Subfamily Cyclosalpinae
        *Helicosalpa virgula* (Vogt, 1854)
        *Cyclosalpa pinnata* (Forskål, 1775)
        *C. affinis* (Chamisso, 1819)
        *C. bakeri* Ritter, 1905
        *C. foxtoni* van Soest, 1974
      Subfamily Salpinae
        *Salpa fusiformis* Cuvier, 1804
        *Thalia democratica* (Forskål, 1775)
        *Ihlea punctata* (Forskål, 1775)
        *Iasis zonaria* (Pallas, 1774)
        *Thetys vagina* Tilesius, 1802
        *Pegea confoederata* (Forskål, 1775)
        *Ritteriella picteti* (Apstein, 1904)
  Order Pyrosomida
    Family Pyrosomidae
      *Pyrosoma* (*Pyrostremma*) *spinosum* Herdman, 1888
      *P.* (*Pyrosoma*) *atlanticum* (Péron, 1804)

## Class **APPENDICULARIA**

## *General biology*

General structure and feeding

Appendicularians are small, free-swimming tunicates with a persistent **tail** and notochord, and with features somewhat reminiscent of the 'tadpole' larval stage of the otherwise sessile ascidians (Fig. 1A, B) – hence the alternative name 'Larvacea'. With the exception of *Kowalewskia* (not found in British waters), there is a permanent, short **endostyle** and a pair of branchial openings, or **gill slits**, which lead from the **pharynx** to the exterior. The digestive tract consists of a **mouth**, pharynx, **oesophagus, buccal glands, stomach** (which may be of one or two lobes), **intestine** and a **rectum** which opens ventrally. There is a ventral **heart** and the **gonads** are large and situated posteriorly (Fig. 2). All species except *Oikopleura dioica* are hermaphrodite.

Of special interest are anterior epidermal glands called **oikoplasts** which secrete the 'house' (Figs. 1B, 3) which plays an essential part in the complex feeding mechanism. There are two types: **Fol's oikoplasts** which are the larger of the two, more anteriorly situated and usually elongated; and **Eisen's oikoplasts** which are more central, smaller and roughly square or round. Their precise position and shape are useful in specific recognition, but they can be extremely difficult to see especially in poorly preserved specimens. Each oikoplast comprises an area of enlarged epidermal cells and therefore can only be seen by histological examination of the outside surface, preferably after staining by quick immersion so that only the epithelium is affected.

The feeding system (Fig. 3), first worked out by Lohmann, is excellently described by Berrill (1950, p. 304) and by Hardy (1956, p. 154). In brief, the **house** is a filter, relatively large to give a large area of filtration, and it acts as a protective covering. There are **filtration grids** over **openings** in its rear which prevent the entry of large particles and water is further passed through a pair of very fine filters within the house. The **pores** in these filters are very small, 0.07–0.24 $\mu$m, but the filter has a ratio of open space to supporting framework of greater than 50 % (Flood, 1978). Currents made by movements of the tail draw water into the system through the filters and out through an **exhalent aperture** at the front. The movements of cilia also help in current formation, and the captured food is trapped in mucus secreted by the endostyle and passed, by cilia, to the mouth. Food consists of minute particles, chiefly nannoplankton and small protozoans. Indeed examination of the gut contents of appendicularians can form a useful method of collection of nannoplankton samples. Water currents can be reversed to eject unwanted particles and to clean the filters.

For use in an emergency or when the filter becomes choked, there is a '**door**' in the house through which the animal can escape and start to secrete a new house, which it can do even at 2–4 hour intervals. The door is normally kept closed. Because of this escape mechanism and the very delicate gelatinous

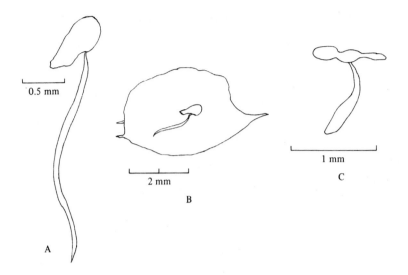

Fig. 1. Outline drawings of appendicularians.
A, *Oikopleura*. B, *Oikopleura* in its house. C, *Fritillaria*.

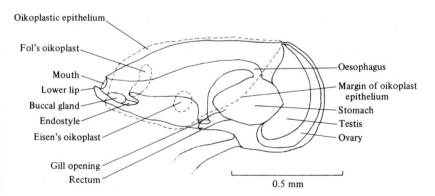

Fig. 2. The general structure of an appendicularian (*Oikopleura*).

structure of the house it is extremely rare to find complete specimens in plankton samples. Although the animals themselves are often abundant, the disintegrating houses form an unrecognizable mush in the sample, sticking to other organisms and making sorting difficult. To see them as a whole necessitates careful collection by water bottle, preferably by a sub-aqua diver. Giant houses seen from 'submersibles' in warm water can be as much as 30–100 cm in size (Barham, 1979).

### Life history

Details of the development of *Oikopleura dioica* were given by Delsman (1910, 1912). The **eggs** are about 0.1 mm in diameter and on dividing gastrulate between the fifth and sixth stages to form a small **tadpole larva** rather similar to that of the ascidians. Various authors have considered the Appendicularia to be very primitive tunicates or neotenic ascidian larvae, whilst Garstang (1928) considered that they had evolved indirectly by way of the Doliolida (see p. 15), although still by way of neotenic larvae, the larval tail being retained and the cloaca lost, the doliolid test then becoming the appendicularian house.

### Locomotion

Movement of the tail within the house draws water through the posterior window in the house and out through the anterior exhalent aperture as part of the feeding mechanism (see above) and this will cause a certain amount of locomotion through the sea, but appendicularians in their houses are drifting, planktonic organisms and cannot really be said to swim. When they escape and become free from the encumbrance of their house, they can swim by a tadpole-like movement but this slight movement is irrelevant to their position in the sea as a whole. Further details of locomotion are given by Bone and Mackie (1975).

### Ecology

Appendicularians are of great importance in marine food chains. Being small and without rapid movement, they are easy prey for carnivorous members of the zooplankton (jellyfish, siphonophores, chaetognaths and lantern-fishes) and they are of particular importance as food for larval and post-larval fish, including the plaice (Shelbourne, 1953, 1962; Ryland, 1964). Their transparent and delicate structure results in their being easily disintegrated and therefore very difficult to recognize in stomach contents, yet their own stomach contents will remain intact for longer. Many reports of unidentifiable green remains, taken to indicate that some phytoplankton or faecal pellets have been taken as food, are likely to be based on the gut contents of appendicularians which have themselves been digested. Appendicularians may well be much more significant as food than the literature suggests.

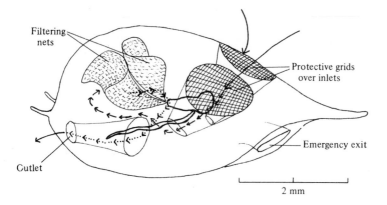

Fig. 3. The feeding mechanism of *Oikopleura* (adapted from Hardy, 1956, with the permission of author and publisher).

## Distribution

Appendicularians are found, often in abundance, from the Poles to the Equator, and with the usual greater diversity of species in warm waters. The abundant species indigenous to British waters are *Oikopleura dioica* and *Fritillaria borealis*, which may be found in British estuaries and inshore waters as well as in the open sea. The fairly tolerant cold-water species *Oikopleura labradoriensis* is frequent in northern British waters, e.g. in the northern North Sea and the Faroe–Shetland Channel, but is rather rare in the southern North Sea and is unlikely to be found inshore. Other species are only occasional visitors to British waters. Associated with influxes of oceanic water are the warm-water species *Oikopleura fusiformis* and *Fritillaria pellucida* and, very rarely indeed, *Oikopleura parva*, *O. albicans*, *O. cophocerea*, *O. longicaudata*, *O. rufescens*, *Althoffia tumida*, *Appendicularia sicula*, *Fritillaria tenella*, *F. gracilis* and *F. aberrans*. The cold-water form *Oikopleura vanhöffeni* has also been recorded in British waters. None of these rare visitors can be considered as belonging to the British fauna and all are absent from British inshore waters: accordingly, they are not considered further here (although those with access to oceanic collections should at least be aware of the possibility of their occurrence).

## Systematic section

There are three families of appendicularians: the Oikopleuridae, Fritillaridae (with two sub-families, the Fritillarinae and Appendicularinae) and Kowalew- skidae. Of these, only the Oikopleuridae and Fritillaridae (Fritillarinae) are represented in British waters. Works which can be recommended for purposes of identification of Appendicularia from wider geographical areas are Büchmann (1969) which treats all species in the North Atlantic and Büchmann and Kapp (1975) which includes all known species.

## Key to British appendicularians

1. Body* ovoid; tail fusiform, narrow at its point of attachment................2

   Body elongated; tail broad, attached near the middle of the body.............. 4

2. Tail with no sub-chordal cells.................................*Oikopleura fusiformis* (p. 12)

   Tail with sub-chordal cells (see Fig. 4)..................................................3

3. Tail with two sub-chordal cells.......................................*Oikopleura dioica* (p. 10)

   Tail with a series of sub-chordal cells........................... *O. labradoriensis* (p. 11)

4. Musculature at tip of tail pointed; end of tail fin square cut or slightly bifurcated.............................. .................... *Fritillaria borealis acuta* (p. 14)

   Musculature at tip of tail square cut; end of tail distinctly bifurcated..............
   .....................................................................................*F. borealis truncata* (p. 14)

---

* Body refers to that of the animal itself, i.e. not including its secreted house.

## Family OIKOPLEURIDAE

The only genus of this family that occurs in British waters is *Oikopleura* Mertens in which both Fol's and Eisen's oikoplasts are present. The anterior (Fol's) is rather crescent-shaped whilst the other (Eisen's), not far behind it, is smaller and roughly a rounded–square shape (Fig. 2). The organs are tightly arranged in a compact body (Fig. 2). The oikoplast epithelium extends dorsally over the coil of the gut, ventrally to the anus. There are gill openings on both sides of the rectum, the posterior part of the stomach is divided into two lobes, and the oesophagus leads into the dorsal edge of the left stomach lobe while the intestine leads from the ventral edge of the right stomach lobe. The tail is several times the body length (Figs. 1A, B; 3).

There are two sub-genera, both the species likely to be found in British waters belonging to *Vexillaria* which has buccal glands on both sides of the endostyle, etchings in that part of the body surface from which the house is secreted (not at all easy to see and quite impossible in badly preserved specimens) and sub-chordal cells in the tail. *O. fusiformis* is a member of the second sub-genus, *Coecaria*, in which there are no buccal glands, no etchings and no sub-chordal cells in the tail.

*Oikopleura dioica* Fol, 1872

(Fig. 4)

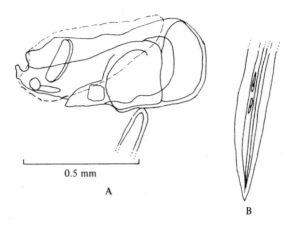

0.5 mm

A

B

Fig. 4. *Oikopleura dioica.*
A, body. B, tail, showing the two sub-chordal cells.

A small species in which the body length is usually between 0.5 and 1.0 mm, although it may reach 1.3 mm. The tail is about four times the body length and has a narrow tail muscle and two distinct sub-chordal cells about one-half to two-thirds down the length of the tail (Fig. 4B). In *O. dioica* the males and females are separate, all other species being hermaphrodite. The buccal glands are small and spherical, and the right stomach lobe forms a cloacal sac behind the entrance to the intestine.

*Oikopleura labradoriensis* Lohmann, 1886

(Fig. 5)

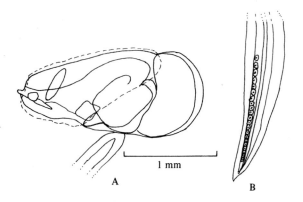

Fig. 5. *Oikopleura labradoriensis.*
A, body. B, tail showing the series of sub-chordal cells.

This is a relatively cold-water species which is often quite abundant in the more northerly British waters. It is larger and more elongate that *O. dioica* with a body length of over 2.0 mm (up to 2.6 mm). There is a long series of sub-chordal cells in the distal third of the tail (Fig. 5b). The left stomach lobe is roughly pentagonal, the ovary is single and the testes are paired. The buccal glands are comparatively large and oval.

*Oikopleura fusiformis* Fol, 1872

(Fig. 6)

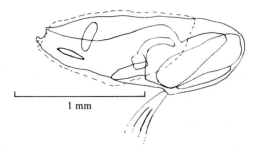

Fig. 6. *Oikopleura fusiformis.*
There are no sub-chordal cells.

*O. fusiformis* is a member of the sub-genus *Coecaria* and therefore lacks buccal glands, sub-chordal cells in the tail and etchings in the germ of the house (see p. 9). Its elongate body attains a length of 1.4 mm. There is a large diverticulum extending from the left stomach and both the intestine and rectum are elongated. The mouth is turned up so that it opens dorsally and possesses a pronounced lower lip. The ovary is rather flat; the paired testes are long, covering most of the stomach.

## Family FRITILLARIDAE

In contrast to the Oikopleuridae, fritillarids have elongate bodies with a broader tail, giving a distinct hammer-shaped appearance (Fig. 1C). Fol's oikoplast is absent and the gill slits open directly to the exterior without tubes. The food trap (house) protrudes freely in front of the mouth when the animal feeds; the anus is on the right.

Only one genus is present in British waters: *Fritillaria* Quoy and Gaimard, 1833. In members of this genus, the organs are fairly free within the body cavity; dorsally, the oikoplast covers only the area above the branchial cavity in which lie the gills, whilst ventrally it forms a narrow zone below the endostyle. The gill openings are close to the endostyle and much in front of the anus. The oesophagus goes straight to the stomach, which is round and composed of only a few cells.

Although Büchmann (1969) quotes seven species and varieties of *Fritillaria* as occurring in off-shore British waters, most are very rare visitors and only *F. borealis* Lohmann, 1896, in both its varieties, is indigenous. Details of the structure and behaviour of the related *F. pellucida* are given by Bone, Gorski and Pulsford (1979).

*Fritillaria borealis* Lohmann, 1896

(Fig. 7)

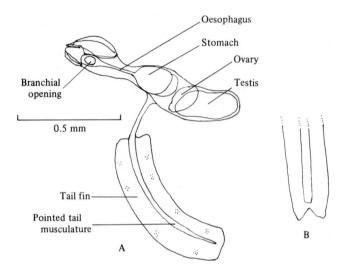

Fig. 7. *Fritillaria borealis.*
A, *Fritillaria borealis acuta.* B, *Fritillaria borealis truncata.*

This species occurs in British waters as two varieties. *F. borealis acuta* Lohmann and Büchmann, 1926 (Fig. 7A) is the cold-water variety. In this, the body length reaches 1.3 mm, the tip of the tail musculature is pointed and the end of the tail fin is rather square cut or slightly bifurcated. The warm-water form, *F. borealis truncata* Lohmann and Büchmann, 1926 (Fig. 7B), is somewhat smaller – body length rarely reaching 0.9 mm – and the tip of its tail musculature is cut roughly square whilst the tail fin is distinctly bifurcated.

## Class **THALIACEA**

Pelagic tunicates with a permanent transparent **test** which can be small or very conspicuous. The musculature consists of loop-like bands around the body, the rhythmic – though not continuously regular – contractions of which draw in water through the anterior aperture and out through the posterior (cloacal) aperture. Alternation of generations occurs. The class comprises three orders, Doliolida, Salpida and Pyrosomida, here treated separately.

## Order **DOLIOLIDA**

Because the loop-like musculature is mostly in the form of complete bands round the body, this order is sometimes also known as the Cyclomyaria. There is a single family, the Doliolidae (Claus, 1882).

## *General biology*

### General structure and feeding

Doliolids are comparatively small, barrel-shaped tunicates rarely exceeding 25 mm in length (Fig. 8). The **test** is thin, making them flabby when lifted out of the water. The **gills** are fairly large, but delicate, usually occupying the anterior half or more of the body. The alimentary canal is ventral and near the posterior end of the gill. There are eight or nine **muscle bands** depending on the stage of the life history (although in trophozooids – see below – they are reduced to about four), the first and last being much narrower than the others and forming sphincters to the entrance and exit apertures (Fig. 8); the

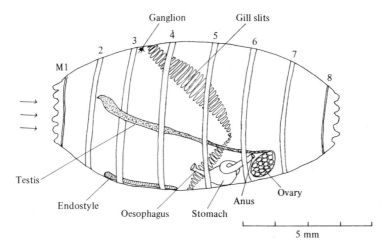

Fig. 8. The general structure of a doliolid (*Doliolum gegenbauri*).

latter may therefore be very easily overlooked when counting the bands.

The alimentary canal comprises a short **oesophagus**, a **stomach** and an **intestine** which may be straight or coiled – a factor useful in taxonomy. Doliolids are hermaphrodite but the reproductive organs are confined to the gonozooid stage (see below). The **ovary** lies ventrally behind the stomach and the **testes** originate near the anterior edge of the ovary and extend forwards, ventrally or on the left side, in various patterns. The positions of the internal organs relative to the muscle bands are particularly useful in identification.

The simplified **endostyle** lies ventral and anterior to the **gill slits** and it secretes mucus which is stretched across the **pharynx** by the action of cilia to form a net which retains minute flagellates, etc. from the water drawn in by the pulsating muscles.

### Life history

Eggs, sexually produced by the **gonozooid** generation, are shed via the **cloacal cavity** and each forms a larva with its characteristically tunicate tadpole-like tail and notochord (Fig. 9C, D). The larva develops inside a membraneous house, from which it eventually becomes free, loses its tail and develops into an **oozooid**, the sexual generation (Fig. 9A). Oozooids have nine muscle bands, an **otolith** on the left side between muscle bands 3 and 4 and a dorsal **nerve**

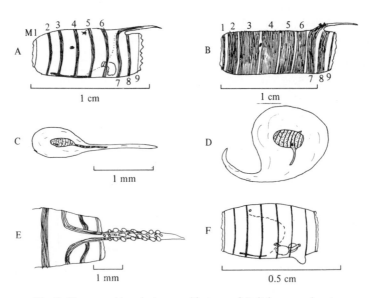

Fig. 9. The oozooid and phorozooid stages of *Doliolum gegenbauri*.
A, the functional oozooid with nine muscle bands, gut, dorsal process (cadophore) and showing the path taken by the migrating buds. B, the old nurse stage. C and D, two early stages of development. E, cadophore with developing buds. F, the sexless phorozooid stage with eight muscle bands and ventral protuberance.

**ganglion** between muscle bands 4 and 5. A dorsal projection – the **cadophore** – pointing rearwards, arises at about muscle band 7 and takes this muscle band, and also to a lesser extent band 8, with it. The gill slits are reduced to four on each side. There is a ventral **stolon** within the parental cuticle which buds off new zooids – **blastozooids** – which migrate along a definite pathway (Fig. 9A) on the right side to the dorsal process. Here they take up position as a central dorsal row flanked on both sides by lateral rows (Fig. 9E). At this stage the oozooid is termed a **nurse**. The nurse serves to produce and carry the developing **buds**, the dorsal cadophore becoming much extended to accommodate them; then the doliolid can be considered a true colony.

The two outer rows of buds on the cadophore develop into **trophozooids** which functionally become the feeding individuals for the whole colony; they are without gonads and have a very reduced musculature. The median row of buds develop into **phorozooids**, which later break away and become free-swimming. They have eight muscle bands but no gonads and they retain the vestige of the attachment stalk as a **ventral process.** Whilst still attached to the cadophore of the nurse, more migratory buds – **pro-buds** – settle on the stalk of the phorozooids and remain there when the phorozooid breaks free. They become elongated and, by further budding, produce gonozooids.

The gonozooids are the free-swimming sexual generation, possessing eight muscle bands and both male and female gonads. The other internal organs are arranged as in the phorozooid although there is no ventral process. As the gonozooids are produced by asexual budding, they are found more abundantly than the other stages.

Having carried the developing buds, the oozooid nurse gradually disintegrates, first losing its internal structure which leaves it empty, although it survives a considerable time as an **old nurse** (Fig. 9B). The otolith falls out as the cuticle becomes fragile and decays, but its loss is not a satisfactory measure of age of the nurse. The muscle bands also gradually spread until in some species they actually meet in one continuous sheet, the **cuirass.** The muscle arrangement in old nurses can be used as a means of identification, and the following terminology applies:

**stenomyonic** – muscle bands narrower than $\frac{1}{2} \times$ the interspaces

**eurymyonic** – muscle bands broader than $\frac{1}{2} \times$ the interspaces

**holomyonic** – muscle bands 2–8, united into a cuirass

**amphiclinous** – muscle bands 2–8, in a series gradually becoming broader and then narrower

**myoplane** – the position in the series from which the muscle bands become narrower both anteriorly and posteriorly

**aclinous** – muscle bands 2–8 not forming an amphiclinous series

Much is now known of the embryology of doliolids, following the work of Godeaux (1955, 1961, 1971) and Braconnot (Braconnot, 1964, 1967; Braconnot and Cassanova, 1967).

## Ecology

That doliolids have an oceanic distribution and so can be used as indicators
of water movement is referred to below. Little positive information is
available on the use of doliolids as food by planktonic predators but it seems
likely that where they are numerous they will be eaten. Because the test is so
delicate – and without it the internal organs and their relationship to each
other will be unrecognizable – many doliolids may well necessarily be
dismissed as 'remains' when examining gut contents. Like appendicularians,
the doliolids may be more important as food than the literature suggests,
although in British waters their numbers are such as to make it unlikely that
their contribution is significant.

## Distribution

Doliolids are oceanic warm-water species and there are no known cold-water
forms (although Harant and Vernières, 1938, stated that *Doliolum intermedia*
and *D. krohni* are from cold water). They are most abundant – both
numerically and in species – in tropical and sub-tropical waters (including the
Mediterranean Sea) and are, therefore, only visitors to British waters brought
in by the warmer ocean currents. Only the more tolerant doliolids survive this
transport successfully and they are therefore good indicators of water
movements in the off-shore rather than in-shore areas. Doliolids only thrive in
the British area in off-shore waters and they are unlikely to survive near
estuaries, even in areas affected by warm-water inflow. There is no known
record of them surviving the winter so as to be found here in the spring.

The most tolerant doliolid is *Doliolum gegenbauri*, both in its *sensu strictu*
form and as forma *tritonis* (see p. 22). The latter is larger and more typical of
the open North Atlantic Ocean, while the former is more characteristic of the
Lusitanian stream bringing water to the west of the British Isles from the
Biscayan area and off the Mediterranean region. In years when the Lusitanian
stream is more than usually in evidence, *D. gegenbauri sensu strictu* can
reproduce *en route* and be carried into the North Sea from the north (Lucas,
1933; Fraser, 1952). Because of the time taken to reach the British Isles,
particularly the northern area, the greatest likelihood of their occurrence is
not the height of summer but from the late summer and autumn, to October.
*Doliolum nationalis* penetrates into the western English Channel (Russell and
Hastings, 1933) and also has once been recorded from west of the Hebrides.

Further offshore, to the south-west of Ireland, four specimens of *Doliolum
mülleri* were taken in 1953 at 47°22 N, 9°30 W (Fraser, 1955A). Other
unlikely migrants into British waters are *Doliolum krohni* and *D. intermedium*.
Harant and Vernières (1938) quoted *D. intermedium* as a cold-water species
occurring in the North Atlantic and North Sea, but this seems very doubtful
indeed. These authors also listed *D. krohni* as a cold-water species but this
seems to have been based on a misidentification of *D. mülleri* var. *krohni*, and
there is no evidence for its being from cold water.

# Identification

The classification of doliolids is based on the internal anatomy of the gonozooid and phorozooid stages, and the free-living stages of doliolids can be separated thus:

With nine muscle bands or a continuous muscle sheet, and with a dorsal cadophore ..................................................................................................................**oozooid**

With eight muscle bands, a vestigeal ventral stalk, but no gonads ...**phorozooid**

With eight muscle bands, no outgrowths from the test, with gonads .................
.................................................................................................................**gonozooids**

The scheme of Garstang (1933), in which the single genus *Doliolum* is divided into four sub-genera, is adopted here. The four sub-genera are:

*Dolioletta* – Alimentary canal forming a close dextral coil in the middle of the cloacal floor and with a median anus (Fig. 10A, B)

*Doliolum* (*sensu strictu*) – Alimentary canal forming a wide dextral arch round the cloacal floor and with a parietal anus on the right side (Fig. 10C)

*Doliolina* – Alimentary canal forming an upright U- or S- shaped loop in the sagittal plane (Fig. 10D)

*Dolioloides* – Alimentary canal extending horizontally in the sagittal plane and ending in a sub-terminal anus (no British species).

Species identification is based on the position and extent of the internal organs relative to the muscle bands. In the following Key and Table 1 these are referred to by number; e.g. M.3 indicates muscle band 3, and $3\frac{1}{4}$ means situated between muscle bands 3 and 4, but nearer to M.3 than M.4. For further details clarifying the older confusion in taxonomy, see Garstang (1933).

Table 1. *The anatomical features of systematic importance of the British doliolids*

| Species | Alimentary Canal | Gill lamella Dorsal attachment | Gill lamella Posterior limit | Gill lamella Ventral attachment | Endostyle Anterior | Endostyle Posterior | Ganglion | Testis | Length (mm) |
|---|---|---|---|---|---|---|---|---|---|
| D. (Dolioletta) gegenbauri Uljanin | Intestine forming a close dextral coil with median anus | $3+$ | $5\frac{3}{4}$ | $5$ | $2\frac{1}{2}-2\frac{3}{4}$ | $4\frac{1}{2}-4\frac{3}{4}$ | $3\frac{1}{4}$ | Extends forwards and obliquely on left side, usually to about M.2, but very variable (M.5 to M.1) and crosses M.3 (and M.2) between the muscle and the ectoderm | 9 |
| D. gegenbauri var. tritonis Herdman | Intestine forming a close dextral coil with median anus | $3+$ | $5\frac{3}{4}$ | $4\frac{1}{2}-4\frac{7}{8}$ | $2\frac{1}{2}-2\frac{3}{4}$ | $4\frac{1}{2}-4\frac{3}{4}$ | $3\frac{1}{4}$ | Extends forwards and obliquely on left side, usually to about M.2, but very variable (M.5 to M.1) and crosses M.3 (and M.2) between the muscle and the ectoderm | 17 |
| D. (Doliolum) nationalis Borgert | With dextral arch, anus on right side about M.6 | $2+$ | $5+$ | $4\frac{1}{2}-5$ | $2+$ | $4+$ | $3\frac{3}{4}$ | Extending horizontally on left side, variable length | $3+$ |
| D. (Doliolina) mülleri Krohn | Upright U-shaped | $5$ | $5\frac{1}{2}$ | $5$ | $2\frac{2}{3}$ | $4\frac{2}{3}$ | $3\frac{1}{4}$ | Forming a ventral hernia between M.5 and M.6 | 4 |

After Fraser (1947).

## Systematic section

## Key to British doliolids

1. Intestine forming a close dextral coil with a median anus; testis (if present) extending forward and obliquely to the left.................................................... **2**

   Intestine with a dextral arch, anus on right side; testis (if present) extending slightly forward in the median plane........................... *D. nationalis* (p. 23)

   Intestine upright and U-shaped; testis (if present) short, between M.5 and M.6 and partly in a ventral hernia ...........................................*D. mülleri* (p. 23)

2. Ventral attachment of gill lamella at M.5...........................*D. gegenbauri* (p. 22)

   Ventral attachment of gill lamella just in front of M.5................................................ ............................................................. *D. gegenbauri* var. *tritonis* (p. 22)

*Doliolum (Dolioletta) gegenbauri* Uljanin, 1884

(Figs. 8, 9, 10A, B)

This is the most frequent of the doliolids in British waters. It is a fairly small species, reaching to about 9 mm in length, with a coiled intestine and a median anus. The testis is very characteristic, reaching forwards and upwards in an oblique line usually to about M.2 but the position and shape of its forward extremity is very variable indeed (it may only reach M.5 but may extend right forward to M.1). As the testis extends forward in the body cavity it grows more and more to the left and presses forwards between M.3 and the body wall and also left (outside) of M.2 when it passes that muscle band. Numerous pairs of gills are attached anteriorly and dorsally just behind M.3, extending rearwards almost to M.6 before turning forwards to their lower attachment at M.5. The endostyle extends from forwards of M.3 almost to M.5; the nerve ganglion is just behind M.3. There are twelve lobes to the fringe at the anterior aperture and eight at the posterior.

D. (*Dolioletta*) *gegenbauri* var. *tritonis* Herdman, 1888 (Fig. 10B) is a larger and more oceanic form than *D. gegenbauri sensu strictu*, reaching up to 17 mm in length. Apart from size, the main distinction of this variety is that the gill slits extend further forward and are attached just in front of M.5.

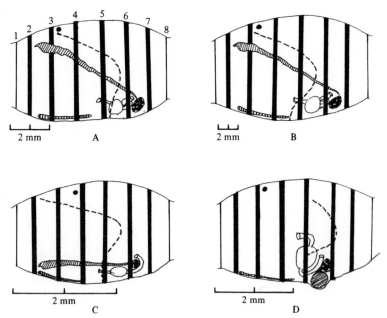

Fig. 10. Comparative diagrams of the gonozooid stage of the British doliolids. A, *Doliolum gegenbauri*. B, *D. gegenbauri* var *tritonis*. C, *D. nationalis*. D, *D. mülleri*.

### *Doliolum (Doliolum) nationalis* Borgert, 1894

#### (Fig. 10C)

A small species, about 3 mm in length, likely to be found in British waters in or near the western entrance of the English Channel. The intestine forms a dextral arch, with the anus on the right side at about M.6. Dorsally, the gill slits are attached behind M.2, extend backwards to M.5, and then turn slightly forward to their lower attachment just in front of M.5. The endostyle starts at about M.2 and extends to M.4; the nerve ganglion is just in front of M.4. The testis is not oblique as in *D. gegenbauri*, but extends horizontally on the left side; its length is variable.

### *Doliolum (Doliolina) mülleri* Krohn, 1852

#### (Fig. 10D)

Only rarely taken in off-shore waters near the British Isles, *D. mülleri* is small (about 4 mm in length) and has an upright U-shaped intestine. The gill slits are less extensive than in the other British species, being attached both dorsally and ventrally at M.5 and curving only slightly to the rear in between. The testis is ball-shaped and lies in a ventral pocket between M.5 and M.6. The endostyle starts in front of M.3 and ends in front of M.5; the nerve ganglion is just in front of M.4.

## Order SALPIDA

Generally speaking, salps are more noticeable in the plankton than the doliolids, as they are usually larger and have a much greater proportion of jelly. The muscle bands may or may not anastomose, but they are discontinuous ventrally (except in *Helicosalpa* and in some muscles of the aggregate form of *Thalia*) and hence the alternative name of Desmomyaria. The pattern of the muscle bands is one of the main features used in identification. There is a single family, the Salpidae.

Alternation of generations occurs although in a much simpler form than in the Doliolida. The sexual and asexual forms are so different (Fig. 11) that until the complete life histories were known, earlier taxonomists regarded them as different species and accordingly gave them separate names. Early works on salps, although often of excellent quality, can be misleading in this respect. Thompson (1948) provides an excellent, simple review of the early nomenclature and taxonomic history of this group. Since then, there has been much work done on the taxonomy and relationships of salps (see Yount, 1954). The asexual forms live singly and are called the **solitary** phase. They produce a **stolon** which buds into a chain of sexual individuals which adhere together for a time and are therefore termed the **aggregate** phase.

As in the Doliolida there are no indigenous British species: those found in British waters have been brought in from warm or western oceanic areas and they do not survive the winter. Salps are therefore also good indicators of the movement of such waters to near Britain (Russell, 1935; Fraser, 1949, 1969).

## *General biology*

### General structure and feeding

Salps are roughly cylindrical in shape but with numerous variations (fusiform, prismatic, dorso-ventrally flattened, etc.) and often with various patterns of posterior projections. Like doliolids, an anterior (**oral**) and posterior (**cloacal**) **aperture** are present. The series of muscle bands can be reduced in some species (e.g. *Pegea*) whilst in others there up to twenty or more. The number of fibres in each muscle band varies, being larger in the higher temperate latitudes and in sub-Antarctic waters and lower in the tropics (van Soest, 1979). In addition, there are smaller muscle bands controlling the oral and cloacal apertures.

The contraction of these muscles in turn draws water through the oral aperture and pushes it out through the cloacal, giving the salp a forward movement and in so doing providing a respiratory current which also contains food organisms from the plankton. The **oral** and **atrial muscles** control the oral and cloacal apertures so ensuring the passage of water in one direction only. The food comprises micro-organisms, chiefly microflagellates and small diatoms. Salps appear to be non-selective feeders and different species occurring together take the same food items (Silver, 1975).

A single gill bar (**branchial tube**) is present, attached dorsally, and more or less anteriorly, and then extending, sloping ventrally, to its posterior and ventral attachment (Fig. 11A). The **endostyle** is ventral and runs almost the length of the salp. A **dorsal nerve ganglion** is present and possesses an outgrowth in the form of a horseshoe-shaped 'eye', the structure of which is much the same in all solitary individuals, but in the aggregate forms the eyes are of different shapes and can be of taxonomic value: they have no optical function. Near the anterior dorsal side is situated the **ciliated funnel** which is also of various patterns and thus of taxonomic value. The ciliated funnel is at the rear of the oral aperture and aids in the collection of food for passage down the **oesophagus** to the **stomach**. The cloaca opens into the **cloacal chamber** which is separated from the **pharyngeal chamber** by a septum.

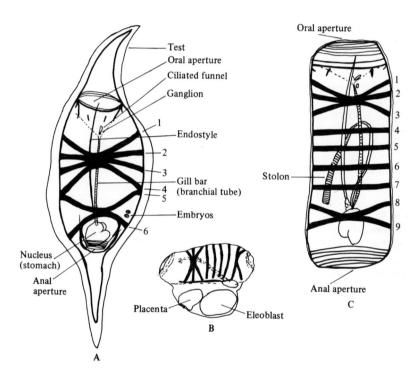

Fig. 11. The general structure of a salp (*Salpa fusiformis*).
A, aggregate form (<100 mm). B, embryo. C, solitary form (<115 mm).

Some of the cyclosalps (sub-family Cyclosalpinae – see p. 28) possess **luminous organs**, usually situated laterally or dorso-laterally. Cyclosalps recorded from British waters possess the following numbers of these organs:

|  | solitary form | aggregate form |
|---|---|---|
| *Cyclosalpa affinis* | nil | nil |
| *C. bakeri* | 5 pairs (rarely 4) | nil |
| *C. pinnata* | 5 pairs (rarely 4) | 1 pair |
| *Helicosalpa virgula* | 1 pair | nil |

### Life history

In the sexual form, the single ovary is on the right side of the body while the male organs are beside the intestine. After fertilization the young embryos (Fig. 11B) grow attached to the integument inside the cloacal chamber until of sufficient size to be released as free-swimming asexual forms. These then, in turn, produce a **stolon** (Fig. 11C) which consists of a chain or whorls of sexual forms, thereby forming a clear alternation of generations in a much simpler form than in the doliolids. The asexual (solitary) and sexual (aggregate) phases are very different in shape and structure, but the relationships are quite clearly seen because the asexual embryos develop inside the body of the sexual phase for long enough for their characteristics to be seen. The young sexual phases budding on the stolon can also be identified at an early stage and long before they break away. As the chains become longer and longer, they break into groups, often associated with their age on the stolon, and do not become separated into individuals until broken by wave action or other disturbance. As this can be so variable, the size of the individuals breaking away can also be very variable. Each solitary form produces a chain of many individuals whilst the aggregate forms produce very few individuals: thus in a thriving community of salps there are usually far greater numbers of the aggregate forms. Unlike the doliolids, the older solitary phases do not disintegrate into 'old nurses', but retain their function and structure until final disintegration or death.

Most of the solitary forms are more or less symmetrical, but the aggregate phases of some cyclosalps (*Cyclosalpa bakeri* and *Helicosalpa virgula*) are markedly asymmetrical, the asymmetry being produced by the method of budding on the stolon. The aggregate phase of *Ihlea punctata* is also asymmetrical.

Much detail of the histology and early life history of a cyclosalp is given by Brooks (1876), of *Salpa fusiformis* by Sutton (1960) and Sawicki (1966), and of salps in general by Ihle (1935).

### Ecology

Salps have an important role in the economy of the sea mainly because of their capacity for existing in immense swarms which can, by grazing, severely reduce the food available to other herbivores in the plankton. This in turn reduces the food available to carnivorous zooplankton and developing young fish (Fraser, 1962). After being drifted into British waters from off-shore regions, salps disintegrate and thus release from their bodies into local waters the end products from phytoplanktonic productivity often from great distances away. This could result in greater quantities of phytoplankton in British waters and therefore also of a greater zooplanktonic production the following spring, and so lead, for example, to an improvement in Shetland herring fisheries (Fraser, 1963).

There are very few reports of fish and other carnivores feeding on salps. Because of the relatively large proportion of jelly, they are presumably not particularly attractive to eat. As salps often exist in swarms to the detriment of other planktonic organisms, however, predators may then need to take them (see Thompson, 1948; Yount, 1958, p. 128). Cod and herring are both known to have done so; and mackerel (in Japanese waters), trigger fish, file fish and Antarctic birds have also been reported to feed on salps (see also Silver, 1975; Godeaux, 1977). It may also be that fish known to feed on ctenophores would also take salps; and the chaetognath *Sagitta* will feed on salps (Oliver, 1954). The insides of salps are eaten out by the amphipod *Phronima* which then utilizes the gelatinous test as a protective house (see Hardy, 1956). Salps with firm tests are obviously more durable for this purpose than those with softer tests, and Fraser (1961) quotes definite identifications of the test of *Iasis zonaria*. Salps living in warm waters are often hosts of ectoparasitic copepods, especially of *Sapphirina* spp. (Heron, 1973).

Salps may impinge directly on man. Beklemishev (1958) wrote a small paper entitled 'Plankton stops a ship' in which he gave details of a very large swarm of *Thalia longicaudata* at 21°S, 7°E distributed in lanes 1 m wide, 1–10 m apart and 3 km in length. They were abundant enough to stop the ship, not by actually hindering progress but by choking the filters of the cooling water intake. Brattström (1972) also reported that the 1955 swarm of salps caused a serious setback to Norwegian fisheries. The general ecology and distribution of salps are discussed in further detail by Yount (1958).

## Systematic section

There are two main groups of salp: the *Cyclosalpa* group in which the stomach is not consolidated into a single ball or close curve (the **nucleus**) and the budding on the stolon is either in whorls (*Cyclosalpa*) or linear (*Helicosalpa*); and the *Salpa* group in which the stomach is consolidated into a nucleus and the stolon buds in linear form.

Metcalfe (1918) separated the cyclosalps into two categories, 'Cyclosalpa asymmetricales' (including *Cyclosalpa bakeri* and *Helicosalpa virgula*) and 'Cyclosalpa symmetricales' (including *Cyclosalpa floridana*, *C. pinnata* and *C. affinis*). The two groups were together known by Metcalfe as the Dolichodea; Yount (1954) grouped them as the sub-family Cyclosalpinae. Metcalfe also introduced the terms Sphaerodea for those salps with a compact stomach nucleus, Circodea for those with a curved stomach and Oligomyaria for those Circodea with a reduced musculature. These terms are now rarely used.

Current practice is to apportion the salps to two sub-families of the single family Salpidae: the Cyclosalpinae and the Salpinae. Van Soest (1974*a*) has reviewed the taxonomy of the world's cyclosalps, including many more species and sub-species than were known to Metcalfe. Two genera are recognized. *Cyclosalpa* in which the budding on the stolon forming the sexual phases, is in a whorl and the individuals (except in *C. bakeri*) are bilaterally symmetrical, and *Helicosalpa* in which the sexual forms are produced in a chain of alternate right- and left-sided forms so that individuals are not symmetrical. Van Soest (1974*a*) also provides charts of the world-wide distribution of the Cyclosalpinae. *Helicosalpa virgula* and *Cyclosalpa bakeri* are known from British waters (Fraser, 1961) – although van Soest (1974*a*) considers that the record of *C. bakeri* may be in error for *C. foxtoni*, a very similar species not described until 1974 – and van Soest considers that *C. pinnata* and *C. affinis* might be expected to occur. *C. pinnata* has, so far, been found only in Atlantic waters north of 30°N.

Salps other than the Cyclosalpinae form the sub-family Salpinae in which the numerous sub-genera of the old genus *Salpa* have been up-graded into genera (see Yount, 1954). The sub-family now comprises eleven genera – *Brooksia*, *Ritteriella*, *Metcalfina*, *Iasis*, *Thetys*, *Thalia*, *Pegea*, *Traustedtia*, *Salpa*, *Weelia* and *Ihlea* – several of which are monotypic. There are many tropical and sub-tropical species and subspecies that have not yet been recorded as occurring as visitors to British waters; some of these may turn up in the future.

# Key to British salps

As the chief features used in identification are the numbers, position and anastomosing of the muscle bands, this key should be used in conjunction with Figs. 11–22 in which the species found in or near British waters are figured in semi-diagrammatic form to emphasize these features.

Stolon present (Fig. 11C)............................................................solitary (asexual) forms

Stolon absent (Fig. 11A)...........................................................aggregate (sexual) forms

# Key to aggregate forms

Note: these may be single individuals, in chains or in whorls.

1. Stomach not concentrated into a compact ball or close curve .......................
...................................................................................................(Cyclosalpinae) **2**

   Stomach concentrated into a compact ball or close curve ......... (Salpinae) **6**

2. Body muscle descending into the stalk by which the individuals are joined together in the whorl (the 'peduncle')........................................ (Cyclosalpa) **3**

   No body muscles descending into the peduncle *Helicosalpa virgula* (p. 31)

3. Body muscles arranged symmetrically.................................................................**4**

   Body muscles arranged asymmetrically.................................................................**5**

4. Luminous organs present ..........................................*Cyclosalpa pinnata* (p. 32)

   Luminous organs absent................................................*Cyclosalpa affinis* (p. 33)

5. Dorsal tubercle a deeply concave C-shape; total number of fibres in muscle bands 1–4, 22–29.....................................................*Cyclosalpa bakeri* (p. 34)

   Dorsal tubercle a shallow C-shape; total number of fibres in muscle bands 1–4, 18–26.........................................................................*Cyclosalpa foxtoni* (p. 35)

6. Body very distinctly fusiform (Fig. 11A)................................................................**7**

   Body slightly fusiform (Fig. 19A) ........................................................................**8**

   Body not at all fusiform (Fig. 21A)........................................................................**10**

7. Musculature symmetrical on ventral side (Fig. 11A).........................................
...................................................................................... *Salpa fusiformis* (p. 36)

   Musculature asymmetrical on ventral side (Fig. 18B).......................................
.................................................................................*Ihlea punctata* (p. 39)

8. Test very firm, posterior end markedly pointed (Fig. 19A) ...............................
.............................................................................................*Iasis zonaria* (p. 41)

   Test flabby .......................................................................................................**9**

9. Body slightly pointed anteriorly; muscle band 1 divides into two (Fig. 22A)
..........................................................................................*Ritteriella picteti* (p. 44)

   Body not pointed anteriorly; muscle band 1 does not divide into two (Fig. 17A)...........................................................................*Thalia democratica* (p. 37)

10. Muscle bands 3 and 4 unite centrally to form a cross (Fig. 21A) ....................
..................................................................*Pegea confoederata* (p. 43)

Muscle bands 3 and 4 distinctly separate (Fig. 20A) ...............................
.................................................................. *Thetys vagina* (p. 42)

## Key to solitary forms

1. Stomach not concentrated into a compact ball or close curve ......................
.............................................................. **2** (Cyclosalpinae)

   Stomach concentrated into a compact ball or close curve ..........**6** (Salpinae)

2. Ventral longitudinal muscle absent................................................**3** (*Cyclosalpa*)

   Ventral longitudinal muscle present ...................... *Helicosalpa virgula* (p. 31)

3. Dorsal longitudinal muscles absent ................................................ **4**

   Two dorsal longitudinal muscles present ................................ **5**

4. Luminous organs absent............................................ *Cyclosalpa affinis* (p. 33)

   Luminous organs present ...........................................*Cyclosalpa pinnata* (p. 32)

5. Five pairs of luminous organs situated between the muscles ...........................
...............................................................*Cyclosalpa bakeri* (p. 34)

   Three or four pairs of luminous organs situated on the muscles....................
...............................................................*Cyclosalpa foxtoni* (p. 35)

6. Body muscles less than 15 ........................................................... **7**

   Body muscles more than 20......................................................... **11**

7. Body very firm; muscle bands wider than the interspaces.................................
.........................................................................*Iasis zonaria* (p. 41)

   Body not especially firm; muscle bands narrower than the interspaces......**8**

8. Posterior of body with spiky projections (Fig. 17B)..................................
.......................................................... *Thalia democratica* (p. 37)

   Body without distinct appendages........................................................**9**

9. Only four muscle bands, stomach in region of bands 3 and 4...........................
......................................................................*Pegea confoederata* (p. 43)

   More than eight muscle bands (bands 1, 2 and 3 anastomose in mid-dorsal line), stomach in region of bands 8 and 9........................................**10**

10. Nine muscle bands, bands 1, 2 and 3 also 8 and 9 anastomose......................
.................................................................. *Salpa fusiformis* (p. 36)

   More than nine muscle bands, bands 1, 2 and 3 also 9 and 10 anastomose ...........................................................*Ihlea punctata* (p. 39)

11. Small, narrow species with almost parallel sides; muscle bands and interspaces approximately equal (Fig. 22B) ..........*Ritteriella picteti* (p. 44)

   Very large, flabby species; anterior portion wider than posterior; muscle bands very narrow and in places forming a lattice-like structure (Fig. 20C)................................................ *Thetys vagina* (p. 42)

*Helicosalpa virgula* (Vogt, 1854)

(Fig. 12)

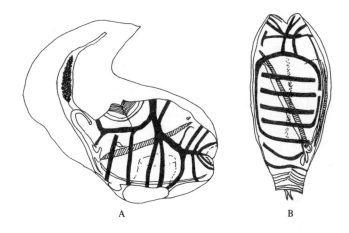

Fig. 12. *Helicosalpa virgula.*
A, aggregate form from right side (<35 mm). B, solitary form (<150 mm).

*Aggregate form* (Fig. 12A). Firm globular test up to about 35 mm in length. A posterior projection, entirely accommodating the elongate testis, is directed dorsally and forward, on the left side in dextral individuals and on the right side in sinistral ones. The muscles are arranged as in Fig. 12A and are markedly asymmetrical.

*Solitary form* (Fig. 12B). The large individuals, up to about 150 mm in length, possess a thick, gelatinous and flabby test. The muscles are arranged as in Fig. 12B. Luminous organs are arranged in a line – continuous in some specimens, somewhat broken in others – from just in front of muscle band 1 to band 6. The dorsal tubercle is highly convoluted in large specimens, but somewhat less so in smaller ones. The stolon does not produce a whorl, but alternate dextral and sinistral individuals. The gut is large and bears two caeca.

Yount (1954) and van Soest (1974*b*) provide further details.

## *Cyclosalpa pinnata* (Forskål, 1775)

(Fig. 13)

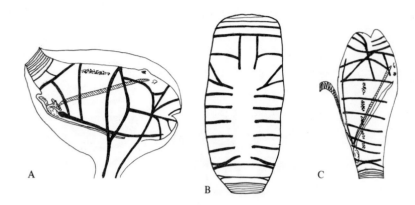

Fig. 13. *Cyclosalpa pinnata.*
A, aggregate form (<64 mm). B, solitary form from above. C, solitary form from side showing stolon and luminous organs (<75 mm).

*Aggregate form* (Fig. 13A). Aggregates are produced in whorls. The test is quite thick and solid and reaches up to about 64 mm in length. Body muscles 1 and 2 fuse dorsally and latero-ventrally before extending into the peduncle (the stalk by which the individuals are joined together in the whorl). Muscle bands 3 and 4 approach dorsally. One strong luminous organ occurs between muscle bands 2 and 3. The intestine bears a single caecum.

*Solitary form* (Fig. 13B). The test is fairly thick and up to about 75 mm in length. The body muscles are all interrupted dorsally. There are five pairs of luminous organs situated between the muscle bands, from band 1 to band 6. The intestine bears two caeca of unequal length. The stolon originates near muscle band 5 and extends forward ventrally to about band 2, whence it projects through the test to the outside, producing the young zooids in whorls.

*Cyclosalpa affinis* (Chamisso, 1819)

(Fig. 14)

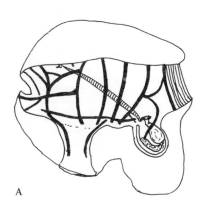

A                                                    B

Fig. 14. *Cyclosalpa affinis.*
A, aggregate form (<80 mm). B, solitary form (<80 mm).

*Aggregate form* (Fig. 14A). The test, up to about 80 mm in length, is thin and soft, and possesses a prominent ventral swelling containing the intestine. Luminous organs are absent; the peduncle is short. Muscle bands 2 and 3 are in continuous loops, whilst band 1 has branches extending forwards mid-laterally and band 4 is similarly branched to the posterior.

*Solitary form* (Fig. 14B). The thick, solid test may attain a length of up to 80 mm. All main muscle bands are interrupted ventrally; bands 1 and 2 are also interrupted in the mid-dorsal line, and other muscles are variably so. Muscle band 1, as in the aggregate form, has forwardly extending branches laterally. Luminous organs are absent. The stolon extends from about muscle band 6 to band 2, whence it projects through the test to the exterior, the attached zooids being formed in whorls.

*Cyclosalpa bakeri* Ritter, 1905

(Fig. 15)

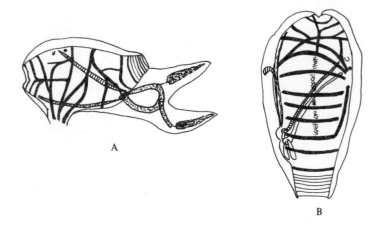

Fig. 15. *Cyclosalpa bakeri.*
A, aggregate form (<50 mm). B, solitary form (<50 mm).

*Aggregate form* (Fig. 15A). The test is thin and flabby, distinctly asymmetrical (with both dextral and sinistral forms) and possesses two posterior projections. Two main muscle groups occur, one of bands 1 and 2, the other of bands 3 and 4; these are not asymmetrical. The gut is in the form of a posterior loop at the base of the two posterior projections; a caecum and a 'problematic' organ (possibly blood forming) extend into the left projection, whilst the long testis occupies the right projection. Luminous organs are absent. The peduncle is moderately long – about one-third of the body length.

*Solitary form* (Fig. 15B). The thin, flabby and rather glutinous test is up to about 47 mm in length. The muscle bands are interrupted dorsally and ventrally (excepting band 7 which may be continuous dorsally). Muscle band 6 extends forwards dorsally as far as band 3 in the form of two parallel longitudinal muscles. Five well-developed pairs of luminous organs occur between the muscle bands and a smaller pair are located in front of muscle band 1 (Yount, 1954, states that the luminous organs are on the muscle bands but it is likely that his specimens were of the closely-similar *C. foxtoni*). The intestine bears two equal and fairly short caeca. The stolon extends from muscle band 5 to band 2 and bears the developing aggregate forms in whorls of up to nine individuals.

*Cyclosalpa foxtoni* van Soest, 1974

(Fig. 16)

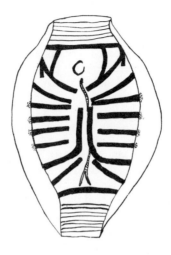

Fig. 16. *Cyclosalpa foxtoni*, solitary form (<37 mm).

*Aggregate form.* This phase of *C. foxtoni* has yet to be completely described, but it is expected to be similar to that of *C. bakeri* but with fewer muscle fibres and a shallower (less concave) dorsal tubercle.

*Solitary form* (Fig. 16). This form is also similar to that of *C. bakeri* but with fewer muscle fibres and with the luminous organs actually on muscle bands 2–5 rather than between them. The dorsal tubercle is a smooth C-shape. The test is rather globular and reaches some 40 mm in length.

*C. foxtoni* has not yet been definitely identified from British waters.

*Salpa fusiformis* Cuvier, 1804

(Fig. 11)

*Aggregate form* (Fig. 11A). The aggregate form is markedly fusiform in shape and bears symmetrical muscle bands both dorsally and ventrally. Muscle bands 1, 2 and 3–4 fuse or approach very closely mid-dorsally, bands 5 and 6 also fuse mid-dorsally, whilst bands 4 and 5 touch laterally. All muscle bands are interrupted ventrally. The stomach is in the form of a very compact nucleus which in life is sage green.

*Solitary form* (Fig. 11C). The shape of the solitary phase is not fusiform and in the early literature was described as *S. runcinata*. Two rather indefinite posterior projections occur. Muscle bands 1, 2 and 3 fuse in the mid-dorsal line; bands 4–7 are separate; and bands 8 and 9 also fuse mid-dorsally. All muscles are interrupted ventrally. The stolon is distinct and extends forward to between muscle bands 4 and 5, and then turns posteriorly. The gut is in the form of a compact nucleus which in life is sage green with streaks of red.

*S. fusiformis* is by far the most abundant salp to be found in British waters. It is brought in from the open ocean to the west almost every summer and is carried north and east towards Shetland and the Norwegian Sea. There is a marked year by year variation in the numbers brought in, and in the extent of penetration into the North Sea from the north (Fraser, 1969); in some years, e.g. 1932–1937, it hardly penetrated at all, while in 1946 it reached as far south as Flamborough Head (54°N). Swarms are occasionally found near St Andrews, Fife, and near the Farne Islands. It occurs in the Celtic Sea and western end of the English Channel, rarely in the eastern end and it does not penetrate into the southern North Sea via the Straits of Dover. It is also quite rare at the northern and southern entrances to the Irish Sea and is not recorded from its central part.

Its greatest abundance in the open ocean is at the height of summer and because it takes time to be carried from there, it is found most commonly in British waters in September and October.

The somewhat spiny form, *S. fusiformis aspersa*, has not yet been identified from British waters and neither have a number of closely related species, e.g. *S. thompsoni* and *S. foxtoni* (see Foxton, 1961).

*Thalia democratica* (Forskål, 1775)

(Fig. 17)

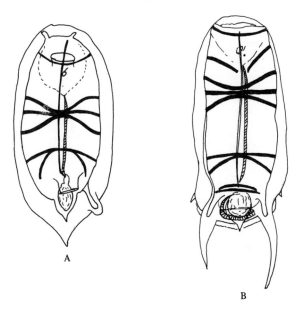

Fig. 17. *Thalia democratica.*
A, aggregate form (<15 mm). B, solitary form (<25 mm).

*Aggregate form* (Fig. 17A). Although the posterior is somewhat fusiform in shape, the anterior end is rounded. Muscle bands 1, 2 and 3 fuse mid-dorsally, as do bands 4 and 5 (5 is sometimes considered to be a rather poorly developed branch of 4), but these two groups are quite separate laterally (in contrast to *S. fusiformis*). The stomach is not quite so tightly packed as in *S. fusiformis*; it is situated in the posterior projection. *Thalia* is slightly asymmetrical in that the atrial opening is not central and there is a posterior lateral protuberance on one side only.

*Solitary form* (Fig. 17B). In the early literature, this was known as *Salpa mucronata*. The posterior end bears numerous, rather spiny, not quite symmetrical projections, similar to but less marked than those possessed by warm-water species of *Thalia*. Muscle bands 1–4 are continuous dorsally and ventrally; band 5 is narrowly interrupted ventrally; bands 1–3 converged mid-dorsally, as do bands 4 and 5, whilst bands 3 and 4 converge ventrally. The stomach partly extends into the middle posterior projection where the short stolon curves round it without projecting forwards.

*Thalia democratica* is one of the world's most abundant salps. It often occurs in immense swarms, several square kilometres in extent. It is less tolerant to the conditions found as far north as the British Isles than is *S. fusiformis*, and whilst it occurs off the western end of the English Channel and off the south-west of Ireland, it swarms only rarely west of Scotland (Fraser, 1961). Some muscles in the aggregate form may be continuous ventrally.

*Ihlea punctata* (Forskål, 1775)

(Fig. 18)

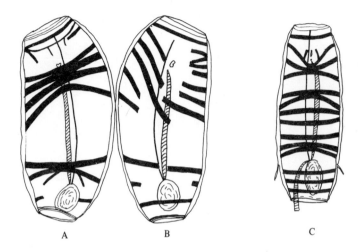

Fig. 18. *Ihlea punctata* (=*I. asymmetrica*).
A, aggregate form, dorsal view (<20 mm). B, aggregate form, ventral view. C, solitary
form (<25 mm).

*Aggregate form* (Fig. 18A, B). The test is very flabby, up to about 20 mm in length, and fusiform, though not as markedly as in *S. fusiformis* which it resembles. Muscle bands 1 and 2 unite mid-dorsally and this pair is very closely joined by bands 3 and 4, making a group of four; bands 5 and 6 also united mid-dorsally. All muscle bands are interrupted ventrally and it is here that the asymmetry is most marked (see Fig. 18B). The stomach is in the form of a nucleus.

*Solitary form* (Fig. 18C). The flabby test is up to about 25 mm in length. This form also resembles the corresponding one of *S. fusiformis*, but, whereas muscle bands 1–3 unite dorsally, bands 4 and 5 closely approach instead of being quite separate; bands 6–8 are separate, whilst bands 9 and 10 approach closely. The stomach forms a compact nucleus. The stolon turns backwards in the region of muscle band 9.

Yount (1954) considers that *I. punctata* and *I. asymmetrica* are the same species and, following this, the recordings of *I. asymmetrica* in British waters should be referred to *I. punctata*. It is a species with a very delicate test which easily disintegrates and it may therefore be more common in British waters than records suggest. It occurs singly or in far less abundant swarms than *Salpa* or *Thalia*, but over much the same area as *S. fusiformis*. Often only the disintegrating nucleus is found, sometimes with barely enough loosely-attached muscle remains to confirm identity. Viewed dorsally the species is symmetrical, but ventrally it is markedly asymmetric in the aggregate form.

*Iasis zonaria* (Pallas, 1774)

(Fig. 19)

Fig. 19. *Iasis zonaria.*
A, aggregate form (<60 mm). B, solitary form (<50 mm). C, cross-section of solitary
form.

*Aggregate form* (Fig. 19A). The very tough, prismatic, and quite large test (up to about 60 mm in length) is asymmetrical posteriorly and deeply grooved dorsally and ventrally; a single, pointed posterior projection is sited off-centre. The muscle bands are very clear, and wider than in most salps; all are interrupted ventrally and band 1 is interrupted dorsally; bands 4 and 5 unite dorsally.

*Solitary form* (Fig. 19B, C). The test, up to about 50 mm in length, is very firm, tough and even more prismatic than in the aggregate form. The muscle bands are extremely wide, almost touching each other laterally; all are interrupted dorsally and ventrally. A very marked, pointed projection is present posteriorly; the stolon forms a circle around the gut.

Because of the tough prismatic test in both solitary and aggregate phases, this species retains its identity very clearly for a long time and is easily identified. It never occurs in large numbers in British waters, but is found chiefly offshore over deep water to the west and it can be carried towards the Norwegian coasts.

*Thetys vagina* Tilesius, 1802

(Fig. 20)

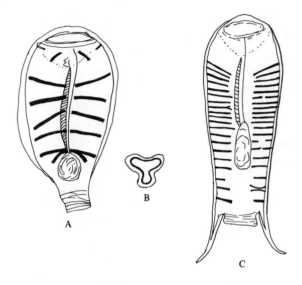

Fig. 20. *Thetys vagina.*
A, aggregate form (<250 mm). B, cross-section of aggregate form. C, solitary form
(<230 mm).

*Aggregate form* (Fig. 20A, B). A very large species, reaching up to some 250 mm length, with a thick, firm test bearing a broad dorsal depression. Five muscle bands are present, each interrupted both dorsally and ventrally; bands 1, 2 and 3 converge dorsally but do not join. The gut is fairly compact and is situated in a ventral swelling of the test.

*Solitary form* (Fig. 20C). Up to about 230 mm in length, the large, thick and firm test is somewhat triangular in section and bears two distinct posterior projections. The muscle bands are numerous, some twenty or more, but weakly developed, often broken in places and interrupted dorsally and ventrally. The fairly compact gut is situated about two-thirds of the body length from the anterior end. The stolon extends forwards before turning posteriorly left of the gut.

To date, only two recordings of this species have been made in British waters, both of the aggregate form. One was a large specimen (235 mm in length) taken in a herring net off Wick in 1929 (Thompson, 1948); the other was collected at 53°22´N, 14°40´W in 1958 (Fraser, 1961).

*Pegea confoederata* (Forskål, 1775)

(Fig. 21)

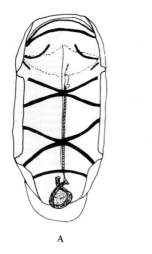 

A                                    B

Fig. 21. *Pegea confoederata.*
A, aggregate form (<150 mm). B, solitary form (<120 mm).

*Aggregate form* (Fig. 21A). This quite large species, up to some 150 mm in length, possesses a loose flabby test. Muscle bands 1 and 2 unite mid-dorsally to form a cross, as do muscle bands 3 and 4; the bands are short, being interrupted before reaching the mid-lateral region.

*Solitary form* (Fig. 21B). The test is quite large, reaching up to 120 mm in length, and pear-shaped, being smaller posteriorly. As in the aggregate form, the four main muscles form two distinct crosses (bands 1 and 2, and 3 and 4). The gut is fairly compact and is encircled closely by the stolon.

Only rarely has *P. confoederata* been recorded from British waters. It was taken in the English Channel in November 1968 (Marine Biological Association Report No. 13) and a single specimen was obtained in November 1958 from 59°24′N, 3°45′W (Fraser, 1961).

*Ritteriella picteti* (Apstein, 1904)

(Fig. 22)

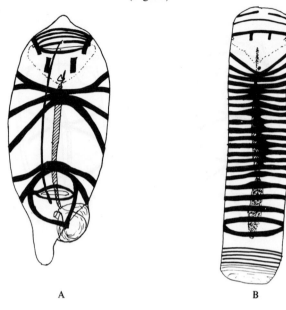

A B

Fig. 22. *Ritteriella picteti.*
A, aggregate form (<25 mm). B, solitary form (<86 mm).

*Aggregate form* (Fig. 22A). This form is slightly fusiform, the flabby test bearing short anterior and posterior projections and being up to 25 mm in length. Muscle bands 1–4 fuse mid-dorsally, as do bands 5 and 6; bands 4 and 5 closely approach laterally. The gut forms a moderately compact ball in the posterior body projection.

*Solitary form* (Fig. 22B). An elongate salp with a length of up to 86 mm, although usually much less. A variable number of muscle bands, 13–24, are present, continuous dorsally and interrupted ventrally; anteriorly, the first five or six are connected by muscle tissue, and the rest are variably connected. The gut is in the form of a somewhat sinuous sausage in the region of the posterior six muscle bands. The stolon turns towards the posterior in the region of muscle bands 5–8.

HMS *Challenger* took seven aggregate forms and one solitary form of this species from the south-west of Ireland in April 1953 (at 47°31′N, 14°21′W). It had not previously been found so far north (Fraser, 1952, 1955b), nor have any further records been made from waters around the British Isles.

## Order PYROSOMIDA

Pyrosomes are pelagic colonial tunicates now usually regarded as a group of thaliaceans but which nevertheless have affinities with the Ascidiacea. The colony is a tubular structure, closed at one end and open at the other, in which the zooids are arranged each with its oral aperture on the outside of the colony and the anal aperture inside. This arrangement results in water being sucked into the colony from all over its outer surface, passed into the central tube, and ejected from the single open end, giving rise to a slow but continuous type of jet propulsion. Colonies are transparent, usually colourless but sometimes bluish, yellowish or slightly grey in colour. The name derives from the brilliant phosphorescence, a light organ being present on each side of the branchial sac of each zooid.

They are found in the warmer seas of the world, from the surface to some 500 m depth (Thompson, 1948), but they have been taken in hauls from as deep as 4000 m (Herdman, 1888). They are very rare indeed in British waters, and associated only with incoming water movements, although Hardy (1956) states that they never come near the British Isles (but see p. 48).

The single family Pyrosomidae is generally regarded as being mono-generic, the single genus *Pyrosoma* comprising some twenty species and sub-species – it is probable, however, that there is much hybridization and many of the different forms may merely be growth stages (Sewell, 1953). Nevertheless there are two distinct groups of pyrosomes: the 'Pyrosoma fixata' and the 'Pyrosoma ambulata'. Garstang (1928) considered them to be separate genera and gave the name *Pyrostremma* to the 'Pyrosoma fixata'. The difference lies in the method of budding from the **stolon**. In the 'Pyrosoma fixata', the **buds** are formed on the stolon, which originates near the posterior end of the endostyle, and they stay in position to form new zooids in the colony – they therefore tend to be in rows. In the 'Pyrosoma ambulata' (*Pyrosoma sensu strictu*), one zooid is formed at a time and is set free before the next is formed; it then moves to settle at the apertural end of the colony. Colonies of 'Pyrosoma ambulata' tend to be firmer than those of 'Pyrosoma fixata'. Here, *Pyrostremma* and *Pyrosoma sensu strictu* are treated as subgenera of *Pyrosoma*.

## *General biology*

### General structure

The **zooids**, more or less oval in shape, are numerous but independent, being held together in a common gelatinous **test**, which is never smooth on the outer surface (Fig. 23A). Each zooid conforms to the characteristic tunicate body plan (Metcalf and Hopkins, 1919; Sewell, 1953), although – for a Thaliacean – the muscles are weakly developed. The alimentary system comprises a **buccal cavity**, a **branchial sac** in a **branchial chamber**, and posterior to the latter a **stomach** and an **intestine** (which invariably has a sinistral bend as it leaves the stomach) which opens into the **atrial chamber**. The ventral **endostyle** extends nearly the whole length of the branchial chamber. **Pharyngeal bands** pass round the **pharynx** and unite behind the **dorsal nerve ganglion**: the bands meet at a broad angle in *Pyrosoma sensu strictu* but extend backwards in *Pyrostremma*. A number of **branchial bars** extend fore and aft in the walls of the branchial sac and help to support the **gill slits (stigmata)**, and the number of these is helpful in identification. The dorsal nerve ganglion is small but prominent and is situated near the anterior end of the branchial sac.

## Life history

An alternation of generations occurs in the pyrosomids. Asexual reproduction has been referred to briefly whilst discussing the differences between *Pyrostremma* and *Pyrosoma sensu strictu*. Eggs from sexual reproduction grow into rudimentary larvae – the **cythozooid** – which have lost all traces of chordate structure. The cythozooid by asexual budding produces the first larval zooids of a colony – the **ascidiozooids** – and it is at this stage that the young colony is set free. Further development of the colony continues asexually by budding from the stolon of the zooids. Small colonies are usually protogynous and larger colonies protandrous. Later development is described in detail by Joliet (1888), Bonnevie (1896), Neumann (1913), Sewell (1953) and van Soest (1974*b*).

## Ecology

Pyrosomes are often found in fish stomachs in those areas in which they are abundant (Bary, 1960; Cowper, 1960; Thompson, 1948). Due to local currents, colonies of *Pyrosoma* may be arranged in parallel lines, with the colonies all arranged facing the same way, the smaller ones nearer the surface and the larger ones more deeply (Bonnier and Perez, 1902).

There are records of only two species in waters which may perhaps be termed British and both are distinctly rare.

# Systematic section

## Key to British pyrosomes

Zooids tending to be in rows; colonies rather flabby; gill bars about 30 or more ..................................................................................*Pyrosoma spinosum* (p. 48)

Zooids tend to appear haphazardly arranged; colonies rather firm; gill bars about 18 .................................................... *Pyrosoma atlanticum giganteum* (p. 48)

*Pyrosoma (Pyrostremma) spinosum* Herdman, 1888

(Fig. 23B)

This is a fragile species and complete colonies are rarely taken in plankton samples. It can occur in vast numbers in tropical waters. Bonnier and Perez (1902) reported that it can reach up to 4 m in length, but Griffin and Yaldwyn (1970) report colonies from Australian waters reaching to lengths of about 14 m; normally specimens are much smaller, often 2 cm or less.

Each zooid has a circle of tentacles near the oral aperture. There are some 55 gill slits and 30 or more gill bars running longitudinal to the long axis; the rows of stigmata lie at about $10°$ to the longitudinal axis. The intestine is elongate and the anus is well posterior of the stomach. The main luminous organ is small and rather indistinct; there are two others, one on each side of the cloaca.

*P. spinosum* was recorded 95 miles south-west of Ireland by Farran (1907); there have been no subsequent records.

*Pyrosoma (Pyrosoma) atlanticum* (Péron, 1804)

(Fig. 23C)

There are seven subspecies of *P. atlanticum* (see Sewell, 1953) only two of which need to be referred to here: *P. atlanticum giganteum* and *P. a. atlanticum*. The latter is mentioned only to avoid confusion, as, in spite of its name, it seems to be confined to the Pacific; the former is the commonest pyrosome in the Atlantic and the only one at all likely to be found in British waters.

These two subspecies are distinguished by the relative number of gill bars, usually 18 in *P. atlanticum giganteum* and 14 in *P. a. atlanticum*, the number of stigmata being 36 and 30, although these figures are not constant but gradually increase as the zooids increase in size and so increase with age of the colony. All other subspecies are confined to the tropics.

Thompson (1948) notes records of *P. atlanticum giganteum* from the North Sea and English Channel which have since been quoted by other authors, but this pyrosome is not included in the 3rd edition of the *Plymouth Marine Fauna* (Marine Biological Association, 1957) and there have been no records of its occurrence from the Scottish plankton collections since at least 1935.

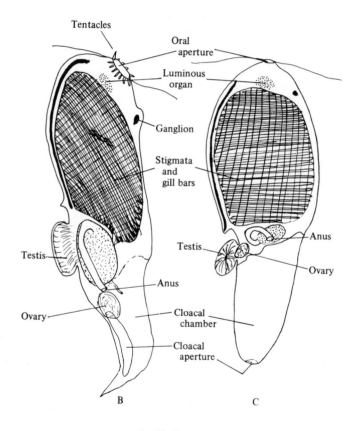

Fig. 23. *Pyrosoma.*
A, general appearance of a pyrosome colony. B, zooid of *Pyrosoma spinosum.*
C, zooid of *Pyrosoma atlanticum giganteum.*

# Acknowledgements

Acknowledgement is gratefully made to the Director of the Marine Laboratory, Aberdeen, to the librarian and his staff and to the typists who have all
done so much to help with the preparation of this work. The figures are
largely based on those of many others, amended or amalgamated by the
author. Fig. 3 is taken, with minor amendments from Fig. 48 in the 'Open Sea'
by permission of Professor Sir Alister Hardy and the publishers Messrs
Collins. My own experiences in the laboratory at Aberdeen and at sea with
research vessels of the Department of Agriculture and Fisheries have also
contributed towards my personal involvement with the pelagic tunicates and
for this I gladly express my gratitude.

# Glossary

**ascidiozooids** The first pyrosome larvae formed by asexual budding and set free to form a new colony.

**atrial chamber** A compartment within the body and into which the cloaca opens.

**atrial muscles** A series of muscle bands, usually thin and sometimes complicated, that control the opening of the atrial chamber to the exterior.

**blastozooid** The early buds, produced asexually, which then develop into the sexual generation.

**branchial bars** The gills are frail structures and may be supported by stronger bands of tissue which are called the branchial bars.

**branchial basket** The structure comprising the gills and the apertures in the gills through which the water passes.

**branchial cavity** The space within the branchial basket.

**branchial openings** The apertures in the gills through which the water passes – see stigmata.

**branchial sac** The pouch formed in front of a series of curved gills.

**branchial tube** The tube through which water passes between the mouth and the gills.

**buccal cavity** The space within the body behind the mouth and from which the oesophagus opens.

**buccal glands** Situated in the oesophagus, near the anterior end of the endostyle, these glands are probably associated with digestion.

**buds** Small developing embryos, before they have formed much internal structure.

**cadophore** A tapering dorsal protuberance in doliolid oozooids which carries the developing embryonic zooids after they have migrated across the body.

**ciliated funnel** A ciliated structure in the buccal cavity; the movements of the cilia help in extracting food particles from the respiratory current and pass them to the oesophagus.

**cuirass** In doliolid oozooids where the muscle bands are so wide that they obliterate the spaces between the bands and so form a continuous muscle sheet.

**cuticle** The outer layer of the body, which gives it its characteristic shape.

**cythozooid** The rudimentary larva of a pyrosome formed by sexual reproduction, and which by asexual budding produces ascidiozooids which form new colonies.

**dorsal nerve ganglion** A small mass of nerve cells situated in the dorsal region, usually in the anterior half of the body.

**endostyle** An elongated glandular structure which produces mucus which helps to trap food particles.

**eye** An organ sensitive to light, but not of optical quality.

**germ of house** Term used by Büchmann (1969) = house surface (Büchmann and Kapp, 1975) = house layer (Thompson, 1948). This refers to the body surface in the region of the oikoplasts which secrete the house.

**gill bar** The relatively rigid structure in salps which supports the delicate gills.

**gonozooid** An individual with sex organs.

**notochord** A primitive type of skeletal rod of turgid cells, the forerunner of the backbone in chordates.

**nucleus** The term used for the compact stomach ball in salps.

**nurse**   The late stage of a doliolid oozooid which carries – or has carried – blastozooids.

**oikoplasts**   Epidermal glands which secrete the gelatinous house in which *Oikopleura* lives.

**oozooid**   An individual which reproduces asexually.

**otolith**   An organ containing a small concretion (statocyst) by which the individual can orientate itself.

**peduncle**   The stalk by which whorls of cyclosalps are attached to their neighbours.

**peripharyngeal bands**   Ciliated bands, in pyrosomes, which encircle the pharynx immediately in front of the branchial cavity.

**phorozooid**   The doliolid stage, incapable of sexual or asexual reproduction, which is largely functional as a means of transport.

**pro-buds**   The developing buds in doliolids which migrate across the body to the cadophore.

**stigmata**   The openings in the gills through which the water passes.

**stolon**   The ribbon-like organ on which the chains, or whorls, of aggregate (sexual) forms are produced by budding.

**sub-chordal cells**   The cells in the tail of appendicularians, lying outside the central core of muscle.

**tentacles**   Filamentous structures, used in this *Synopsis* for those surrounding the oral openings of pyrosomes.

**test**   The gelatinous outer covering of the body.

**trophozooid**   Small rudimentary doliolid zooids whose function appears to be largely concerned with feeding.

**zooid**   The general term applied to individual pelagic tunicates of whatever form or function, occurring either singly or in colonies.

# References

Barham, E. G. (1979). Giant larvacean houses: Observations from deep submersibles. *Science* **205**, 1129–31.

Bary, B. McK. (1960). Notes on Ecology, Distribution and Systematics of Pelagic Tunicata from New Zealand. *Pacific Science* **14**(2), 101–21.

Beklemishev, C. W. (1958). Plankton stops a Ship. *Piroda* 1958(11), 105–6.

Berrill, N. J. (1950). *The Tunicata, with an Account of the British Species*. Ray Society, London, no. 133.

Bone, Q., Gorski, G. and Pulsford, A. (1979). On the Structure and Behaviour of *Fritillaria* (Tunicata: Larvacea). *Journal of the Marine Biological Association UK.* **59**, 399–412.

Bone, Q. and Mackie, G. O. (1975). Skin Impulses and Locomotion in *Oikopleura* (Tunicata: Larvacea). *Biological Bulletin of the Marine Biological Laboratory,* Wood's Hole, Mass. **149**, 267–86.

Bonnevie, K., (1896). On Gemmation in *Distaplia magnilarva* and *Pyrosoma elegans*. *Norwegian North Atlantic Expedition.* 1876–78, *Zoology* 7, 1–16.

Bonnier, J. and Perez, C., (1902). Sur un nouveau Pyrosome Gigantesque. *Comptes Rendus Hebdomadaires des Séances de l'Academie des Sciences, Paris* **134**, 1238–40.

Braconnot, J. C. (1964). Sur le Développement de la Larvae de *Doliolum denticulatum* Q et G. *Comptes Rendus Hebdomadairesdes Séances de l'Academie des Sciences, Paris* **259**, 4361–63.

Braconnot, J. C. (1967). Sur la Possibilité d'un Cycle Court de Développement chez le Tunicier Pelagique: *Doliolum nationalis* Borgert. *Comptes Rendus Hendomadaires des Séances de l'Academie des Sciences, Paris* **264**, 1434–7.

Braconnot, J. C. and Cassanova, J. P. (1967). Sur le Tunicier Pelagique *Doliolum nationalis* Borgert 1893 en Méditerranée Occidentale. *Revue des Travaux de l'institut (Scientifique et Technologique) des Pêches Maritime* 31 (4), 393–402.

Brattström, H. (1972). On *Salpa fusiformis* Cuvier (Thaliacea) in Norwegian Coastal and Offshore Waters. *Sarsia* **48**, 71–90.

Brooks, W. K. (1876). Development of Salpa. *Bulletin of the Museum of Comparative Zoology, Harvard University* 111(14), 291–384.

Büchmann, A. (1969). *Appendicularia*. Fiches d'Identification du Zooplankton, no. 7. Conseil Permanent International pour l'Exploration de la Mer, Copenhagen.

Büchmann, A. (1973). Sorted Samples and Qualitative Counts in Appendicularian Catches. *Marine Biology* 21, 349–53.

Büchmann, A. and Kapp, H. (1975). Taxonomic Characters used for the Distinction of Species of Appendicularia. *Mitteilungen aus den Hamburgischen Zoologischen Museum und Institut* 72, 201–28.

Claus, C. F. W. von. (1882). *Grundzüge der Zoologie* 4th edn., vol. 2, pp. 110–34, Marburg.

Cowper, T. R. (1960). Occurrence of *Pyrosoma* on the Continental Shelf. *Nature, London* 187, 878–9.

Delsman, H. C. (1910). Beiträge zur Entwicklungsgeschichte von *Oikopleura dioica*. *Verhandelingen uit het Rijksinstitut Onderzoeh der Zee* 3, 1–24.

Delsman, H. C. (1912). Weitere Beobachtungen über die Entwicklung von *Oikopleura dioica, Tijdscrift van het Nederlandersche Diekundige Vereeniging, Leiden* 12(2), 199–205.

Farran, G. P. (1907). On the distribution of the Thaliacea and *Pyrosoma* in Irish Waters. *Scientific Investigations. Fishery Board, Ireland* (1906) Appendix (11), 3–17.

Flood, P. R. (1978). Filter Characteristics of Appendicularian Food Catching Nets. *Experientia* **34**, 173–5.

Foxton, P. (1961). *Salpa fusiformis* Cuvier and Related Species. *Discovery Report* no. 32, 1–32.

Fraser, J. H. (1947). *Thaliacea I and II. Doliolidae and Salpidae.* Fiches d'Identification du Zooplankton, nos. 9 and 10. Conseil Permanent International pour l'Exploration de la Mer.

Fraser, J. H. (1949). The distribution of Thaliacea (Salps and Doliolids) in Scottish Waters, 1920–1939. *Scientific Investigation of Fisheries Division. Scottish Home Department*, 1949(1), 44 pp.

Fraser, J. H. (1952). The Chaetognatha and other Zooplankton of the Scottish Area and their Value as Biological Indicators of Hydrographical Conditions. *Marine Research* 1952(2), 52 pp.

Fraser, J. H. (1955a). The Plankton of the Waters Approaching the British Isles in 1953. *Marine Research* 1955(1), 12 pp.

Fraser, J. H. (1955b). The Salp *Ritteriella* of the English coast – a Correction. *Journal of the Marine Biological Association UK*. **34**, 247–8.

Fraser, J. H. (1961). The Oceanic and Bathypelagic Plankton of the North-east Atlantic and its Possible Significance to Fisheries. *Marine Research* **1961** (4), 48 pp.

Fraser, J. H. (1962). The Role of Ctenophores and Salps in Zooplankton Production and standing crop. *Rapport et Procès-verbaux des Reunions. (Conseil Permanent International pour l'Exploration de la Mer)* **153**, 121–3.

Fraser, J. H. (1963). Plankton and the Shetland Herring Fishery. *Rapport et Procès-verbaux des Reunions. (Conseil Permanent International pour l'Exploration de la Mer)* **154**, 175–8.

Fraser, J. H. (1968). The History of Plankton Sampling. *Unesco Monograph of Oceanographic Methodology* **2**(1), 11–18. Unesco Press, Paris.

Fraser, J. H. (1969). The Overflow of Oceanic Plankton to the Shelf Waters of the North-east Atlantic. *Sarsia* **34**, 313–30.

Garstang, W. (1928). The Morphology of the Tunicata, and its Bearings on the Phylogeny of the Chordata. *Quarterly Journal of Microscopical Science* **62**, 51–187.

Garstang, W. (1933). Report on the Tunicata. Part 1. Doliolida. *British Antarctic Terra Nova Expedition*, 1910, Zoology **4**(6), 195–251.

Godeaux, J. (1955). Stades Larvaires du *Doliolum. Bulletin d'l. Academie Royale de Belgique. Classe des Sciences* **41**, 769–87.

Godeaux, J. (1961). L'Oozöide de *Doliolum nationalis* Borg. *Bulletin de la Société Royale des sciences de Liège* **1–2**, 6 pp.

Godeaux, J. (1971). Recherches sur l'Endostyle des Tuniciers. *Rapport et Procès-Verbaux des Réunions, Commission Internationale pour l'Exploration Scientifique de la Mer Mediterranee* **20**(3), 367–8.

Griffin, D. J. G. and Yaldwyn, J. C. (1970). Giant Colonies of Pelagic Tunicates (*Pyrosoma spinosum*) from South-east Australia and New Zealand. *Nature, Lond.* **226**, 464–5.

Harant, H. and Vernières, P. (1938). Tuniciers (2), in *Faune de France*. Lechevalier, Paris.

Hardy, A. C. (1956). *The Open Sea, part 1. The World of Plankton*. Collins, London. 335 pp.

Herdman, W. (1888). Report upon the Tunicata. *Report on the scientific results of the "Challenger" Expedition, 1873–76. Zoology* **27**(3), 166 pp.

Heron, A. C. (1973). A Specialised Predator-Prey Relationship between the Copepod *Sapphirina angusta* and the Pelagic Tunicate *Thalia democratica. Journal of the Marine Biological Association, UK*. **53**, 429–35.

Ihle, J. E. W. (1935). Tunicata. 3. Desmomyaria, in *Handbuch der Zoologie*, **5**(2), 401–532.

Joliet, L. (1888). Observations sur la Blastogénèse et sur la Génération Alternante chez les Salpen et Pyrosomes. *Comptes Rendus Hebdomadaires Séances de l'Academie des Sciences, Paris* **96**, 1676–9.

Lucas, C. E. (1933). Occurrence of *Dolioletta gegenbauri* Uljanin in the North Sea. *Nature, London* **132**, 858.

Marine Biological Association, UK. (1957). *Plymouth Marine Fauna* (3rd edn), 457 pp.

Metcalf, M. M. (1918). The Salpidae: a Taxonomic Study. *Bulletin of the US National Museum* **100**(2), pt. 2, 193.

Metcalf, M. M. and Hopkins, H. S. (1919). Pyrosoma. *Bulletin of the US National Museum* **100**(2) pt. 3, 195–272.

Neumann, G. (1913). Die Pyrosoma und Dolioliden der Deutschen Südpolar Expedition, *Wissenschaftliche Ergebnisse Deutsche Südpolar Expedition* **14**(1), 17–34.

Oliver, M. M. (1954). Sobre la Biologie de los Sagitta de Plancton del Levante Español. *Publicaciones del Instituto de Biologia Aplicada, Barcelona* **16**, 137–48.

Russell, F. S. (1935). On the Value of Certain Plankton Animals as Indicators of Water Movements in the English Channel and North Sea. *Journal of the Marine Biological Association, UK.* **20**, 309–32.

Russell, F. S. and Hastings, A. B. (1933). On the Occurrence of Pelagic Tunicates (Thaliacea) in the Waters of the English Channel off Plymouth. *Journal of the Marine Biological Association, UK* **18**, 635–40.

Ryland, J. S. (1964). The Feeding of Plaice and Sand-eel Larvae in the Southern North Sea. *Journal of the Marine Biological Association, UK* **44**, 343–64.

Sawicki, R. M. (1966). Development of the Stolon in *Salpa fusiformis* Cuvier and *Salpa aspersa* Chamisso. *Discovery Report* **33**, 335–83.

Sewell, R. B. S. (1953). The Pelagic Tunicata. *Scientific Report, John Murray Expedition*, 1933–34, **10**(1), 1–90.

Shelbourne, J. E. (1953). The Feeding Habits of Plaice Post-larvae in Good and Bad Plankton Patches. *Journal of the Marine Biological Association*, UK. **36**, 539–52.

Shelbourne, J. E. (1962). A Predator-prey Relationship for Plaice Larvae Feeding on *Oikopleura*. *Journal of the Marine Biological Association*, UK. **42**, 243–52.

Silver, M. W. (1975). The habitat of *Salpa fusiformis* in the California Current as Defined by Indicator Assemblages. *Limnology and Oceanography, Baltimore* **20**, 230–7.

Sutton, M. F. (1960). The sexual Development of *Salpa fusiformis* Cuvier, *Journal of Embryology and Experimental Morphology* **8**, 268–90.

Van Soest, R. W. M. (1974a). Taxonomy of the Sub-family Cyclosalpinae Yount 1954. (Tunicata, Thaliacea) with Descriptions of Two New Species. *Beaufortia* **22**(288), 17–55.

Van Soest, R. W. M. (1974b) Juvenile Colonies of the Genus *Pyrostremma* Garstang, 1929. (Tunicata, Thaliacea) *Bulletin of the Zoological Museum of the University of Amsterdam* **4**(4), 23–33.

Van Soest, R. W. M. (1979). North-south Diversity. In *Zoogeography and diversity in plankton* (ed. R. W. N. van Soest and A. C. Pierrot-Bults). Bunge Science Publications, Utrecht, 460 pp., see pp. 103–11.

Thompson, H. (1948). *Pelagic Tunicata of Australia*. Commonwealth Council for Scientific and Industrial Research, Melbourne. 1–196.

Van Waard-Pouw, G. and Van Soest, R. W. M. (1973). A Bibliography of Pelagic Tunicata in *Verslagen am technischen Gegevens*. 2. Zoological Museum of the University, Amsterdam. 148 pp. (mimeo).

Yount, J. L. (1954). Taxonomy of the Salpidae (Tunicata) of the Central Pacific. *Pacific Science* **8**(3–4), 276–330.

Yount, J. L. (1958). Distribution and Ecological Aspects of Central Pacific Salpidae (Tunicata). *Pacific Science* **12**, 111–30.

# Index of Families, Subfamilies, Genera and Species

For species and genera the correct names are in *italics*; synonyms are in roman.

*aberrans, Fritillaria* 7
*affinis, Cyclosalpa* 26, 28, 33
*albicans, Oikopleura* 7
*Althoffia tumida* 7
*Appendicularia sicula* 7
asymmetrica, Ihlea 39
*atlanticum, Pyrosoma* 48

*bakeri, Cyclosalpa* 26, 28, 34
*borealis, Fritillaria* 7, 13, 14
Brooksia 28

Circodea 28
*Coecaria* 9
*confoederata, Pegea* 43
*cophocerea, Oikopleura* 7
*Cyclosalpa affinis* 26, 28, 33
*Cyclosalpa bakeri* 26, 28, 34
Cyclosalpa floridana 28
Cyclosalpa foxtoni 35
*Cyclosalpa pinnata* 26, 28, 32

democratica, Thalia 37
dioica Oikopleura 4, 6, 7, 10
Dolioletta 19
*Dolioletta gegenbauri* 18, 20, 22
*Dolioletta gegenbauri* v. *tritonis* 18, 20, 22
Doliolina 19
Doliolina mülleri 18, 20, 23
Doliolina mülleri v. krohni 18
Dolioloides 19
Doliolum 19
Doliolum intermedium 18
Doliolum krohni 18
Doliolum nationalis 18, 20, 23

*floridana, Cyclosalpa* 28, 32
*foxtoni, Cyclosalpa* 35
foxtoni, Salpa 36
FRITILLARIDAE 13
*Fritillaria aberrans* 7
*Fritillaria borealis* 7, 13, 14
*Fritillaria borealis* v. *acuta* 14
*Fritillaria borealis* v. *truncata* 14
*Fritillaria gracilis* 7
*Fritillaria pellucida* 7, 13

*Fritillaria tenella* 7
*fusiformis, Oikopleura* 7, 9, 12
fusiformis, Salpa 26, 36, 37

*gegenbauri, Dolioletta* 18, 20, 22
*gegenbauri, Dolioletta* v. *tritonis* 18, 20, 22
*gracilis, Fritillaria* 7

Helicosalpa 24, 28
*Helicosalpa virgula* 26, 28, 31

Iasis 28
Iasis zonaria 27, 41
Ihlea 28
Ihlea asymmetrica 39
Ihlea punctata 26, 39
intermedium, Doliolum 18

Kowalewskia 4
krohni, Doliolum 18

*labradoriensis, Oikopleura* 7, 11
longicaudata, Oikopleura 7
longicaudata, Thalia 27

Metcalfina 28
mucronata, Salpa 37
*mülleri, Doliolum* 18, 20, 23
*mülleri, Doliolum* v. *krohni* 18

*nationalis, Doliolum* 18, 20, 23

*Oikopleura albicans* 7
*Oikopleura cophocerea* 7
*Oikopleura dioica* 4, 6, 7, 10
*Oikopleura fusiformis* 7, 9, 12
*Oikopleura labradoriensis* 7, 11
*Oikopleura longicaudata* 7
*Oikopleura parva* 7
*Oikopleura rufescens* 7
*Oikopleura vanhöffeni* 7
OIKOPLEURIDAE 9

*parva, Oikopleura* 7
Pegea 28
*Pegea confoederata* 43

*pellucida, Fritillaria* 7, 13
*Phronima* 27
*picteti, Ritteriella* 44
*pinnata, Cyclosulpa* 26, 28, 32
*punctata, Ihlea* 26, 39
*Pyrosoma* 45, 46
*Pyrosoma atlanticum* 48
*Pyrosoma atlanticum* v. *giganteum* 48
*Pyrosoma spinosum* 48
**PYROSOMIDAE** 45
*Pyrostremma* 45, 48

*Ritteriella* 28
*Ritteriella picteti* 44
*rufescens, Oikopleura* 7
runcinata, Salpa 36

*Salpa* 28
*Salpa fusiformis* v. *aspersa* 36
*Salpa foxtoni* 36
*Salpa fusiformis* 26, 36, 37
Salpa mucronata 37
Salpa runcinata 36
*Salpa thompsoni* 36

**SALPINAE** 28
*sicula, Appendicularia* 7
Sphaerodea 28
*spinosum, Pyrosoma* 48

*tenella, Fritillaria* 7
*Thalia* 28
*Thalia democratica* 37
*Thalia longicaudata* 27
*Thetys* 28
*Thetys vagina* 42
*thompsoni, Salpa* 36
*Traustedtia* 28
*truncata, Fritillaria* 14
*tumida, Althoffia* 7

*vagina, Thetys* 42
*vanhöffeni, Oikopleura* 7
*Vexillaria* 9
*virgula, Helicosalpa* 26, 28, 31

*Weelia* 28

*yonaria, Iasis* 27, 41